If found please return this

# MOTORSPORT
# LOG BOOK

to:

AF810834

| RACE EVENT | | DATE |
|---|---|---|
| LOCATION | | |
| ORGANISER | | / / |

| | | | |
|---|---|---|---|
| RACE NUMBER IN ROW | | CURRENT STANDING | |
| WEATHER CONDITION | | TIME OF YEAR | |
| TIRES USED | | RACE START | |
| | | RACE FINISH | |
| CIRCUIT NAME | | RACING CLASS | |

| TYRE PRESSURE | | START POSITION |
|---|---|---|
| FRONT | REAR | |
| | | LAPS COMPLETED |
| | | |
| | | 1ST ROUND AT |
| | | |

| FASTEST TIME | FASTEST LAP |
|---|---|
| | |
| | |
| | |

FURTHER OBSERVATIONS

| | | |
|---|---|---|
| RACE EVENT | | DATE |
| LOCATION | | / / |
| ORGANISER | | |

| | | | |
|---|---|---|---|
| RACE NUMBER IN ROW | | CURRENT STANDING | |
| WEATHER CONDITION | | TIME OF YEAR | |
| TIRES USED | | RACE START | |
| | | RACE FINISH | |
| CIRCUIT NAME | | RACING CLASS | |

| TYRE PRESSURE | | START POSITION |
|---|---|---|
| FRONT | REAR | |
| | | LAPS COMPLETED |
| | | |
| | | 1ST ROUND AT |
| | | |

| FASTEST TIME | FASTEST LAP |
|---|---|
| | |
| | |
| | |

**FURTHER OBSERVATIONS**

| | | |
|---|---|---|
| RACE EVENT | | DATE |
| LOCATION | | |
| ORGANISER | | / / |

| | | | |
|---|---|---|---|
| RACE NUMBER IN ROW | | CURRENT STANDING | |
| WEATHER CONDITION | | TIME OF YEAR | |
| TIRES USED | | RACE START | |
| | | RACE FINISH | |
| CIRCUIT NAME | | RACING CLASS | |

| TYRE PRESSURE | | START POSITION |
|---|---|---|
| FRONT | REAR | |
| | | LAPS COMPLETED |
| | | |
| | | 1ST ROUND AT |
| | | |

| FASTEST TIME | FASTEST LAP |
|---|---|
| | |
| | |
| | |

| FURTHER OBSERVATIONS |
|---|
| |
| |
| |

| | | |
|---|---|---|
| RACE EVENT | | DATE |
| LOCATION | | |
| ORGANISER | | / / |

| | | | |
|---|---|---|---|
| RACE NUMBER IN ROW | | CURRENT STANDING | |
| WEATHER CONDITION | | TIME OF YEAR | |
| TIRES USED | | RACE START | |
| | | RACE FINISH | |
| CIRCUIT NAME | | RACING CLASS | |

| TYRE PRESSURE | |
|---|---|
| FRONT | REAR |
| | |
| | |

START POSITION

LAPS COMPLETED

1ST ROUND AT

| FASTEST TIME | FASTEST LAP |
|---|---|
| | |
| | |
| | |

FURTHER OBSERVATIONS

| | | |
|---|---|---|
| RACE EVENT | | DATE |
| LOCATION | | |
| ORGANISER | | / / |

| | | | |
|---|---|---|---|
| RACE NUMBER IN ROW | | CURRENT STANDING | |
| WEATHER CONDITION | | TIME OF YEAR | |
| TIRES USED | | RACE START | |
| | | RACE FINISH | |
| CIRCUIT NAME | | RACING CLASS | |

| TYRE PRESSURE | | START POSITION |
|---|---|---|
| FRONT | REAR | |
| | | LAPS COMPLETED |
| | | |
| | | 1ST ROUND AT |
| | | |

| FASTEST TIME | FASTEST LAP |
|---|---|
| | |
| | |
| | |

FURTHER OBSERVATIONS

| | | |
|---|---|---|
| RACE EVENT | | DATE |
| LOCATION | | |
| ORGANISER | | / / |

| | | | |
|---|---|---|---|
| RACE NUMBER IN ROW | | CURRENT STANDING | |
| WEATHER CONDITION | ☼ ☁ ❄ ⚡ ≋ | TIME OF YEAR | ❀ ☼ 🍂 ❄ |
| TIRES USED | | RACE START / RACE FINISH | 🕒 |
| CIRCUIT NAME | 🏁 | RACING CLASS | |

| TYRE PRESSURE | |
|---|---|
| FRONT | REAR |
| | |
| | |

START POSITION

LAPS COMPLETED

1ST ROUND AT

| FASTEST TIME | FASTEST LAP |
|---|---|
| | |
| | |
| | |

FURTHER OBSERVATIONS

| RACE EVENT | | DATE |
|---|---|---|
| LOCATION | | |
| ORGANISER | | / / |

| RACE NUMBER IN ROW | | CURRENT STANDING | |
|---|---|---|---|
| WEATHER CONDITION | | TIME OF YEAR | |
| TIRES USED | | RACE START | |
| | | RACE FINISH | |
| CIRCUIT NAME | | RACING CLASS | |

| TYRE PRESSURE | | START POSITION |
|---|---|---|
| FRONT | REAR | |
| | | LAPS COMPLETED |
| | | |
| | | 1ST ROUND AT |
| | | |

| FASTEST TIME | FASTEST LAP |
|---|---|
| | |
| | |
| | |

| FURTHER OBSERVATIONS |
|---|
| |
| |
| |

| | | |
|---|---|---|
| RACE EVENT | | DATE |
| LOCATION | | |
| ORGANISER | | / / |

| | | | |
|---|---|---|---|
| RACE NUMBER IN ROW | | CURRENT STANDING | |
| WEATHER CONDITION | ☼ ☁ ❄ ⛈ ≋ | TIME OF YEAR | 🌷 ☼ 🍂 ❄ |
| TIRES USED | | RACE START<br>RACE FINISH | 🕒 |
| CIRCUIT NAME | 🏁 | RACING CLASS | |

## TYRE PRESSURE

| FRONT | REAR |
|---|---|
| | |
| | |

## START POSITION

## LAPS COMPLETED

## 1ST ROUND AT

## FASTEST TIME

## FASTEST LAP

## FURTHER OBSERVATIONS

| | | |
|---|---|---|
| RACE EVENT | | DATE |
| LOCATION | | |
| ORGANISER | | / / |

| | | | |
|---|---|---|---|
| RACE NUMBER IN ROW | | CURRENT STANDING | |
| WEATHER CONDITION | | TIME OF YEAR | |
| TIRES USED | | RACE START | |
| | | RACE FINISH | |
| CIRCUIT NAME | | RACING CLASS | |

| TYRE PRESSURE | | START POSITION |
|---|---|---|
| FRONT | REAR | |
| | | LAPS COMPLETED |
| | | |
| | | 1ST ROUND AT |
| | | |

| FASTEST TIME | FASTEST LAP |
|---|---|
| | |
| | |
| | |

FURTHER OBSERVATIONS

| RACE EVENT | | DATE |
|---|---|---|
| LOCATION | | / / |
| ORGANISER | | |

| RACE NUMBER IN ROW | | CURRENT STANDING | |
|---|---|---|---|
| WEATHER CONDITION | | TIME OF YEAR | |
| TIRES USED | | RACE START | |
| | | RACE FINISH | |
| CIRCUIT NAME | | RACING CLASS | |

| TYRE PRESSURE | | START POSITION |
|---|---|---|
| FRONT | REAR | |
| | | LAPS COMPLETED |
| | | |
| | | 1ST ROUND AT |
| | | |

| FASTEST TIME | FASTEST LAP |
|---|---|
| | |
| | |
| | |

| FURTHER OBSERVATIONS |
|---|
| |
| |
| |

| | | |
|---|---|---|
| RACE EVENT | | DATE |
| LOCATION | | |
| ORGANISER | | / / |

| | | | |
|---|---|---|---|
| RACE NUMBER IN ROW | | CURRENT STANDING | |
| WEATHER CONDITION | | TIME OF YEAR | |
| TIRES USED | | RACE START | |
| | | RACE FINISH | |
| CIRCUIT NAME | | RACING CLASS | |

| TYRE PRESSURE | | START POSITION |
|---|---|---|
| FRONT | REAR | |
| | | LAPS COMPLETED |
| | | |
| | | 1ST ROUND AT |
| | | |

| FASTEST TIME | FASTEST LAP |
|---|---|
| | |
| | |
| | |

FURTHER OBSERVATIONS

| RACE EVENT | | DATE |
|---|---|---|
| LOCATION | | |
| ORGANISER | | / / |

| RACE NUMBER IN ROW | | CURRENT STANDING | |
|---|---|---|---|
| WEATHER CONDITION | ☼ ☁ ❄ ⚡ ≋ | TIME OF YEAR | ❄ |
| TIRES USED | | RACE START / RACE FINISH | |
| CIRCUIT NAME | | RACING CLASS | |

| TYRE PRESSURE | |
|---|---|
| FRONT | REAR |
| | |
| | |

START POSITION

LAPS COMPLETED

1ST ROUND AT

| FASTEST TIME | FASTEST LAP |
|---|---|
| | |
| | |
| | |

FURTHER OBSERVATIONS

| | | |
|---|---|---|
| RACE EVENT | | DATE |
| LOCATION | | |
| ORGANISER | | / / |

| | | | |
|---|---|---|---|
| RACE NUMBER IN ROW | | CURRENT STANDING | |
| WEATHER CONDITION | | TIME OF YEAR | |
| TIRES USED | | RACE START | |
| | | RACE FINISH | |
| CIRCUIT NAME | | RACING CLASS | |

## TYRE PRESSURE

| FRONT | REAR |
|---|---|
| | |
| | |

## START POSITION

## LAPS COMPLETED

## 1ST ROUND AT

## FASTEST TIME

## FASTEST LAP

## FURTHER OBSERVATIONS

| RACE EVENT | | DATE |
|---|---|---|
| LOCATION | | / / |
| ORGANISER | | |

| RACE NUMBER IN ROW | | CURRENT STANDING | |
|---|---|---|---|
| WEATHER CONDITION | ☼ ☁ ❄ ⚡ ≋ | TIME OF YEAR | 🌷 ☼ 🍂 ❄ |
| TIRES USED | | RACE START / RACE FINISH | 🕒 |
| CIRCUIT NAME | 🏁 | RACING CLASS | |

| TYRE PRESSURE | | START POSITION |
|---|---|---|
| FRONT | REAR | |
| | | LAPS COMPLETED |
| | | |
| | | 1ST ROUND AT |
| | | |

| FASTEST TIME | FASTEST LAP |
|---|---|
| | |
| | |
| | |

FURTHER OBSERVATIONS

| RACE EVENT | | DATE |
|---|---|---|
| LOCATION | | |
| ORGANISER | | / / |

| RACE NUMBER IN ROW | | CURRENT STANDING | |
|---|---|---|---|
| WEATHER CONDITION | | TIME OF YEAR | |
| TIRES USED | | RACE START / RACE FINISH | |
| CIRCUIT NAME | | RACING CLASS | |

| TYRE PRESSURE | |
|---|---|
| FRONT | REAR |
| | |
| | |

START POSITION

LAPS COMPLETED

1ST ROUND AT

| FASTEST TIME | FASTEST LAP |
|---|---|
| | |

FURTHER OBSERVATIONS

| RACE EVENT | | DATE |
|---|---|---|
| LOCATION | | |
| ORGANISER | | / / |

| RACE NUMBER IN ROW | | CURRENT STANDING | |
|---|---|---|---|
| WEATHER CONDITION | | TIME OF YEAR | |
| TIRES USED | | RACE START / RACE FINISH | |
| CIRCUIT NAME | | RACING CLASS | |

**TYRE PRESSURE**

| FRONT | REAR |
|---|---|
| | |
| | |

**START POSITION**

**LAPS COMPLETED**

**1ST ROUND AT**

**FASTEST TIME**

**FASTEST LAP**

**FURTHER OBSERVATIONS**

| | | |
|---|---|---|
| RACE EVENT | | DATE |
| LOCATION | | |
| ORGANISER | | / / |

| | | | |
|---|---|---|---|
| RACE NUMBER IN ROW | | CURRENT STANDING | |
| WEATHER CONDITION | | TIME OF YEAR | |
| TIRES USED | | RACE START | |
| | | RACE FINISH | |
| CIRCUIT NAME | | RACING CLASS | |

| TYRE PRESSURE | |
|---|---|
| FRONT | REAR |
| | |
| | |

START POSITION

LAPS COMPLETED

1ST ROUND AT

| FASTEST TIME | FASTEST LAP |
|---|---|
| | |

FURTHER OBSERVATIONS

| | | |
|---|---|---|
| RACE EVENT | | DATE |
| LOCATION | | |
| ORGANISER | | / / |

| | | | |
|---|---|---|---|
| RACE NUMBER IN ROW | | CURRENT STANDING | |
| WEATHER CONDITION | | TIME OF YEAR | |
| TIRES USED | | RACE START / RACE FINISH | |
| CIRCUIT NAME | | RACING CLASS | |

| TYRE PRESSURE | |
|---|---|
| FRONT | REAR |
| | |
| | |

START POSITION

LAPS COMPLETED

1ST ROUND AT

| FASTEST TIME | FASTEST LAP |
|---|---|
| | |

FURTHER OBSERVATIONS

| | | |
|---|---|---|
| RACE EVENT | | DATE |
| LOCATION | | |
| ORGANISER | | / / |

| | | | |
|---|---|---|---|
| RACE NUMBER IN ROW | | CURRENT STANDING | |
| WEATHER CONDITION | | TIME OF YEAR | |
| TIRES USED | | RACE START | |
| | | RACE FINISH | |
| CIRCUIT NAME | | RACING CLASS | |

| TYRE PRESSURE | | START POSITION |
|---|---|---|
| FRONT | REAR | |
| | | LAPS COMPLETED |
| | | |
| | | 1ST ROUND AT |
| | | |

| FASTEST TIME | FASTEST LAP |
|---|---|
| | |
| | |
| | |

FURTHER OBSERVATIONS

| RACE EVENT | | DATE |
|---|---|---|
| LOCATION | | / / |
| ORGANISER | | |

| RACE NUMBER IN ROW | | CURRENT STANDING | |
|---|---|---|---|
| WEATHER CONDITION | ☼ ☁ ❄ ⛈ ≋ | TIME OF YEAR | 🌷 ☼ 🍂 ❄ |
| TIRES USED | | RACE START / RACE FINISH | 🕒 |
| CIRCUIT NAME | 🏁 | RACING CLASS | |

| TYRE PRESSURE | |
|---|---|
| FRONT | REAR |
| | |
| | |

START POSITION

LAPS COMPLETED

1ST ROUND AT

| FASTEST TIME | FASTEST LAP |
|---|---|
| | |
| | |
| | |

FURTHER OBSERVATIONS

| RACE EVENT | | DATE |
|---|---|---|
| LOCATION | | |
| ORGANISER | | / / |

| | | | |
|---|---|---|---|
| RACE NUMBER IN ROW | | CURRENT STANDING | |
| WEATHER CONDITION | | TIME OF YEAR | |
| TIRES USED | | RACE START | |
| | | RACE FINISH | |
| CIRCUIT NAME | | RACING CLASS | |

| TYRE PRESSURE | | START POSITION |
|---|---|---|
| FRONT | REAR | |
| | | LAPS COMPLETED |
| | | |
| | | 1ST ROUND AT |
| | | |

| FASTEST TIME | FASTEST LAP |
|---|---|
| | |
| | |
| | |

| FURTHER OBSERVATIONS |
|---|
| |
| |
| |

| | | |
|---|---|---|
| RACE EVENT | | DATE |
| LOCATION | | |
| ORGANISER | | / / |

| | | | |
|---|---|---|---|
| RACE NUMBER IN ROW | | CURRENT STANDING | |
| WEATHER CONDITION | ☼ ☁ ❄ ⛈ ≋ | TIME OF YEAR | ☼ ❄ |
| TIRES USED | | RACE START | |
| | | RACE FINISH | |
| CIRCUIT NAME | | RACING CLASS | |

| TYRE PRESSURE | |
|---|---|
| FRONT | REAR |
| | |
| | |

START POSITION

LAPS COMPLETED

1ST ROUND AT

| FASTEST TIME | FASTEST LAP |
|---|---|
| | |
| | |
| | |

FURTHER OBSERVATIONS

| RACE EVENT | | DATE |
|---|---|---|
| LOCATION | | |
| ORGANISER | | / / |

| RACE NUMBER IN ROW | | CURRENT STANDING | |
|---|---|---|---|
| WEATHER CONDITION | | TIME OF YEAR | |
| TIRES USED | | RACE START / RACE FINISH | |
| CIRCUIT NAME | | RACING CLASS | |

| TYRE PRESSURE | |
|---|---|
| FRONT | REAR |
| | |
| | |

START POSITION

LAPS COMPLETED

1ST ROUND AT

| FASTEST TIME | FASTEST LAP |
|---|---|
| | |
| | |
| | |

FURTHER OBSERVATIONS

| | | |
|---|---|---|
| RACE EVENT | | DATE |
| LOCATION | | |
| ORGANISER | | / / |

| | | | |
|---|---|---|---|
| RACE NUMBER IN ROW | | CURRENT STANDING | |
| WEATHER CONDITION | | TIME OF YEAR | |
| TIRES USED | | RACE START | |
| | | RACE FINISH | |
| CIRCUIT NAME | | RACING CLASS | |

## TYRE PRESSURE

| FRONT | REAR |
|---|---|
| | |
| | |

## START POSITION

## LAPS COMPLETED

## 1ST ROUND AT

## FASTEST TIME

## FASTEST LAP

## FURTHER OBSERVATIONS

| | | |
|---|---|---|
| RACE EVENT | | DATE |
| LOCATION | | |
| ORGANISER | | / / |

| | | | |
|---|---|---|---|
| RACE NUMBER IN ROW | | CURRENT STANDING | |
| WEATHER CONDITION | | TIME OF YEAR | |
| TIRES USED | | RACE START | |
| | | RACE FINISH | |
| CIRCUIT NAME | | RACING CLASS | |

**TYRE PRESSURE**

| FRONT | REAR |
|---|---|
| | |
| | |

**START POSITION**

**LAPS COMPLETED**

**1ST ROUND AT**

**FASTEST TIME**

**FASTEST LAP**

**FURTHER OBSERVATIONS**

| | | |
|---|---|---|
| RACE EVENT | | DATE |
| LOCATION | | |
| ORGANISER | | / / |

| | | | |
|---|---|---|---|
| RACE NUMBER IN ROW | | CURRENT STANDING | |
| WEATHER CONDITION | | TIME OF YEAR | |
| TIRES USED | | RACE START | |
| | | RACE FINISH | |
| CIRCUIT NAME | | RACING CLASS | |

| TYRE PRESSURE | |
|---|---|
| FRONT | REAR |
| | |
| | |

START POSITION

LAPS COMPLETED

1ST ROUND AT

| FASTEST TIME | FASTEST LAP |
|---|---|
| | |
| | |
| | |

FURTHER OBSERVATIONS

| | | |
|---|---|---|
| RACE EVENT | | DATE |
| LOCATION | | |
| ORGANISER | | / / |

| | | | |
|---|---|---|---|
| RACE NUMBER IN ROW | | CURRENT STANDING | |
| WEATHER CONDITION | | TIME OF YEAR | |
| TIRES USED | | RACE START | |
| | | RACE FINISH | |
| CIRCUIT NAME | | RACING CLASS | |

## TYRE PRESSURE

| FRONT | REAR |
|---|---|
| | |
| | |

## START POSITION

## LAPS COMPLETED

## 1ST ROUND AT

## FASTEST TIME

## FASTEST LAP

## FURTHER OBSERVATIONS

| | | |
|---|---|---|
| RACE EVENT | | DATE |
| LOCATION | | |
| ORGANISER | | / / |

| | | | |
|---|---|---|---|
| RACE NUMBER IN ROW | | CURRENT STANDING | |
| WEATHER CONDITION | | TIME OF YEAR | |
| TIRES USED | | RACE START | |
| | | RACE FINISH | |
| CIRCUIT NAME | | RACING CLASS | |

| TYRE PRESSURE | | START POSITION |
|---|---|---|
| FRONT | REAR | |
| | | LAPS COMPLETED |
| | | |
| | | 1ST ROUND AT |
| | | |

| FASTEST TIME | FASTEST LAP |
|---|---|
| | |
| | |
| | |

FURTHER OBSERVATIONS

| | | |
|---|---|---|
| RACE EVENT | | DATE |
| LOCATION | | |
| ORGANISER | | / / |

| | | | |
|---|---|---|---|
| RACE NUMBER IN ROW | | CURRENT STANDING | |
| WEATHER CONDITION | | TIME OF YEAR | |
| TIRES USED | | RACE START | |
| | | RACE FINISH | |
| CIRCUIT NAME | | RACING CLASS | |

| TYRE PRESSURE | | START POSITION |
|---|---|---|
| FRONT | REAR | |
| | | LAPS COMPLETED |
| | | |
| | | 1ST ROUND AT |
| | | |

| FASTEST TIME | FASTEST LAP |
|---|---|
| | |
| | |
| | |

| FURTHER OBSERVATIONS |
|---|
| |
| |
| |

| | | |
|---|---|---|
| RACE EVENT | | DATE |
| LOCATION | | / / |
| ORGANISER | | |

| | | | |
|---|---|---|---|
| RACE NUMBER IN ROW | | CURRENT STANDING | |
| WEATHER CONDITION | ☼ ☁ ❄ ⛈ ≋ | TIME OF YEAR | 🌷 ☼ 🍂 ❄ |
| TIRES USED | | RACE START / RACE FINISH | 🕒 |
| CIRCUIT NAME | 🏁 | RACING CLASS | |

| TYRE PRESSURE | |
|---|---|
| FRONT | REAR |
| | |
| | |

START POSITION

LAPS COMPLETED

1ST ROUND AT

| FASTEST TIME | FASTEST LAP |
|---|---|
| | |
| | |
| | |

FURTHER OBSERVATIONS

| RACE EVENT | | DATE |
|---|---|---|
| LOCATION | | |
| ORGANISER | | / / |

| RACE NUMBER IN ROW | | CURRENT STANDING | |
|---|---|---|---|
| WEATHER CONDITION | | TIME OF YEAR | |
| TIRES USED | | RACE START / RACE FINISH | |
| CIRCUIT NAME | | RACING CLASS | |

| TYRE PRESSURE | |
|---|---|
| FRONT | REAR |
| | |
| | |

START POSITION

LAPS COMPLETED

1ST ROUND AT

| FASTEST TIME | FASTEST LAP |
|---|---|
| | |

FURTHER OBSERVATIONS

| RACE EVENT | | DATE |
|---|---|---|
| LOCATION | | |
| ORGANISER | | / / |

| RACE NUMBER IN ROW | | CURRENT STANDING | |
|---|---|---|---|
| WEATHER CONDITION | | TIME OF YEAR | |
| TIRES USED | | RACE START / RACE FINISH | |
| CIRCUIT NAME | | RACING CLASS | |

| TYRE PRESSURE | |
|---|---|
| FRONT | REAR |
| | |
| | |

START POSITION

LAPS COMPLETED

1ST ROUND AT

| FASTEST TIME | FASTEST LAP |
|---|---|
| | |
| | |
| | |

FURTHER OBSERVATIONS

| RACE EVENT | | DATE |
|---|---|---|
| LOCATION | | / / |
| ORGANISER | | |

| RACE NUMBER IN ROW | | CURRENT STANDING | |
|---|---|---|---|
| WEATHER CONDITION | | TIME OF YEAR | |
| TIRES USED | | RACE START / RACE FINISH | |
| CIRCUIT NAME | | RACING CLASS | |

| TYRE PRESSURE | |
|---|---|
| FRONT | REAR |
| | |
| | |

START POSITION

LAPS COMPLETED

1ST ROUND AT

FASTEST TIME

FASTEST LAP

FURTHER OBSERVATIONS

| | | |
|---|---|---|
| RACE EVENT | | DATE |
| LOCATION | | |
| ORGANISER | | / / |

| | | | |
|---|---|---|---|
| RACE NUMBER IN ROW | | CURRENT STANDING | |
| WEATHER CONDITION | | TIME OF YEAR | |
| TIRES USED | | RACE START | |
| | | RACE FINISH | |
| CIRCUIT NAME | | RACING CLASS | |

| TYRE PRESSURE | | START POSITION |
|---|---|---|
| FRONT | REAR | |
| | | LAPS COMPLETED |
| | | |
| | | 1ST ROUND AT |
| | | |

| FASTEST TIME | FASTEST LAP |
|---|---|
| | |
| | |
| | |

## FURTHER OBSERVATIONS

| RACE EVENT | | DATE |
|---|---|---|
| LOCATION | | |
| ORGANISER | | / / |

| RACE NUMBER IN ROW | | CURRENT STANDING | |
|---|---|---|---|
| WEATHER CONDITION | | TIME OF YEAR | |
| TIRES USED | | RACE START | |
| | | RACE FINISH | |
| CIRCUIT NAME | | RACING CLASS | |

| TYRE PRESSURE | |
|---|---|
| FRONT | REAR |
| | |
| | |

START POSITION

LAPS COMPLETED

1ST ROUND AT

| FASTEST TIME | FASTEST LAP |
|---|---|
| | |
| | |
| | |

FURTHER OBSERVATIONS

| RACE EVENT | | DATE |
|---|---|---|
| LOCATION | | / / |
| ORGANISER | | |

| RACE NUMBER IN ROW | | CURRENT STANDING | |
|---|---|---|---|
| WEATHER CONDITION | | TIME OF YEAR | |
| TIRES USED | | RACE START / RACE FINISH | |
| CIRCUIT NAME | | RACING CLASS | |

| TYRE PRESSURE | |
|---|---|
| FRONT | REAR |
| | |
| | |

START POSITION

LAPS COMPLETED

1ST ROUND AT

| FASTEST TIME | FASTEST LAP |
|---|---|
| | |

FURTHER OBSERVATIONS

| | | |
|---|---|---|
| RACE EVENT | | DATE |
| LOCATION | | |
| ORGANISER | | / / |

| | | | |
|---|---|---|---|
| RACE NUMBER IN ROW | | CURRENT STANDING | |
| WEATHER CONDITION | | TIME OF YEAR | |
| TIRES USED | | RACE START | |
| | | RACE FINISH | |
| CIRCUIT NAME | | RACING CLASS | |

| TYRE PRESSURE | |
|---|---|
| FRONT | REAR |
| | |
| | |

START POSITION

LAPS COMPLETED

1ST ROUND AT

| FASTEST TIME | FASTEST LAP |
|---|---|
| | |
| | |
| | |

FURTHER OBSERVATIONS

| | | | |
|---|---|---|---|
| RACE EVENT | | DATE | |
| LOCATION | | / / | |
| ORGANISER | | | |
| RACE NUMBER IN ROW | | CURRENT STANDING | |
| WEATHER CONDITION | ☼ ☁ ❄ ⛈ ≋ | TIME OF YEAR | ❀ ☼ ❦ ❄ |
| TIRES USED | | RACE START / RACE FINISH | ◷ |
| CIRCUIT NAME | ⚑ | RACING CLASS | |

| TYRE PRESSURE | | START POSITION |
|---|---|---|
| FRONT | REAR | |
| | | LAPS COMPLETED |
| | | |
| | | 1ST ROUND AT |
| | | |

| FASTEST TIME | FASTEST LAP |
|---|---|
| | |
| | |
| | |

FURTHER OBSERVATIONS

| | | DATE |
|---|---|---|
| RACE EVENT | | |
| LOCATION | | / / |
| ORGANISER | | |

| | | | |
|---|---|---|---|
| RACE NUMBER IN ROW | | CURRENT STANDING | |
| WEATHER CONDITION | ☼ ☁ ❄ ⛈ ≋ | TIME OF YEAR | 🌷 ☼ 🍂 ❄ |
| TIRES USED | | RACE START / RACE FINISH | 🕒 |
| CIRCUIT NAME | 🏁 | RACING CLASS | |

## TYRE PRESSURE

| FRONT | REAR |
|---|---|
| | |
| | |

## START POSITION

## LAPS COMPLETED

## 1ST ROUND AT

## FASTEST TIME

## FASTEST LAP

## FURTHER OBSERVATIONS

| RACE EVENT | | DATE |
|---|---|---|
| LOCATION | | |
| ORGANISER | | / / |

| RACE NUMBER IN ROW | | CURRENT STANDING | |
|---|---|---|---|
| WEATHER CONDITION | | TIME OF YEAR | |
| TIRES USED | | RACE START | |
| | | RACE FINISH | |
| CIRCUIT NAME | | RACING CLASS | |

| TYRE PRESSURE | | START POSITION |
|---|---|---|
| FRONT | REAR | |
| | | LAPS COMPLETED |
| | | |
| | | 1ST ROUND AT |
| | | |

| FASTEST TIME | FASTEST LAP |
|---|---|
| | |
| | |
| | |

| FURTHER OBSERVATIONS |
|---|
| |
| |
| |

| | | |
|---|---|---|
| RACE EVENT | | DATE |
| LOCATION | | / / |
| ORGANISER | | |

| | | | |
|---|---|---|---|
| RACE NUMBER IN ROW | | CURRENT STANDING | |
| WEATHER CONDITION | ☼ ☁ ❄ ⛈ ≋ | TIME OF YEAR | 🌷 ☼ 🍂 ❄ |
| TIRES USED | | RACE START<br>RACE FINISH | 🕒 |
| CIRCUIT NAME | 🏁 | RACING CLASS | |

| TYRE PRESSURE | | START POSITION |
|---|---|---|
| FRONT | REAR | |
| | | LAPS COMPLETED |
| | | |
| | | 1ST ROUND AT |
| | | |

| FASTEST TIME | FASTEST LAP |
|---|---|
| | |
| | |
| | |

FURTHER OBSERVATIONS

| | | |
|---|---|---|
| RACE EVENT | | DATE |
| LOCATION | | |
| ORGANISER | | / / |

| | | | |
|---|---|---|---|
| RACE NUMBER IN ROW | | CURRENT STANDING | |
| WEATHER CONDITION | | TIME OF YEAR | |
| TIRES USED | | RACE START | |
| | | RACE FINISH | |
| CIRCUIT NAME | | RACING CLASS | |

| TYRE PRESSURE | | START POSITION |
|---|---|---|
| FRONT | REAR | |
| | | LAPS COMPLETED |
| | | |
| | | 1ST ROUND AT |
| | | |

| FASTEST TIME | FASTEST LAP |
|---|---|
| | |
| | |
| | |

FURTHER OBSERVATIONS

| RACE EVENT | | DATE |
|---|---|---|
| LOCATION | | |
| ORGANISER | | / / |

| RACE NUMBER IN ROW | | CURRENT STANDING | |
|---|---|---|---|
| WEATHER CONDITION | | TIME OF YEAR | |
| TIRES USED | | RACE START / RACE FINISH | |
| CIRCUIT NAME | | RACING CLASS | |

| TYRE PRESSURE | | START POSITION |
|---|---|---|
| FRONT | REAR | |
| | | LAPS COMPLETED |
| | | |
| | | 1ST ROUND AT |
| | | |

| FASTEST TIME | FASTEST LAP |
|---|---|
| | |
| | |
| | |

FURTHER OBSERVATIONS

| RACE EVENT | | DATE |
|---|---|---|
| LOCATION | | |
| ORGANISER | | / / |

| RACE NUMBER IN ROW | | CURRENT STANDING | |
|---|---|---|---|
| WEATHER CONDITION | | TIME OF YEAR | |
| TIRES USED | | RACE START<br>RACE FINISH | |
| CIRCUIT NAME | | RACING CLASS | |

| TYRE PRESSURE | | START POSITION |
|---|---|---|
| FRONT | REAR | |
| | | LAPS COMPLETED |
| | | |
| | | 1ST ROUND AT |
| | | |

| FASTEST TIME | FASTEST LAP |
|---|---|
| | |
| | |
| | |

| FURTHER OBSERVATIONS |
|---|
| |
| |
| |

| | | |
|---|---|---|
| RACE EVENT | | DATE |
| LOCATION | | / / |
| ORGANISER | | |

| | | | |
|---|---|---|---|
| RACE NUMBER IN ROW | | CURRENT STANDING | |
| WEATHER CONDITION | ☼ ☁ ❄ ⛈ ≋ | TIME OF YEAR | 🌷 ☼ 🍂 ❄ |
| TIRES USED | | RACE START | |
| | | RACE FINISH | |
| CIRCUIT NAME | | RACING CLASS | |

| TYRE PRESSURE | |
|---|---|
| FRONT | REAR |
| | |
| | |

START POSITION

LAPS COMPLETED

1ST ROUND AT

| FASTEST TIME | FASTEST LAP |
|---|---|
| | |
| | |
| | |

FURTHER OBSERVATIONS

| RACE EVENT | | DATE |
|---|---|---|
| LOCATION | | |
| ORGANISER | | / / |

| RACE NUMBER IN ROW | | CURRENT STANDING | |
|---|---|---|---|
| WEATHER CONDITION | ☀ ☁ ❄ ⛈ 〰 | TIME OF YEAR | 🌷 ☀ 🍂 ❄ |
| TIRES USED | | RACE START / RACE FINISH | 🕒 |
| CIRCUIT NAME | 🏁 | RACING CLASS | |

| TYRE PRESSURE | |
|---|---|
| FRONT | REAR |
| | |
| | |

**START POSITION**

**LAPS COMPLETED**

**1ST ROUND AT**

| FASTEST TIME | FASTEST LAP |
|---|---|
| | |
| | |
| | |

**FURTHER OBSERVATIONS**

| RACE EVENT | | DATE |
|---|---|---|
| LOCATION | | |
| ORGANISER | | / / |

| RACE NUMBER IN ROW | | CURRENT STANDING | |
|---|---|---|---|
| WEATHER CONDITION | | TIME OF YEAR | |
| TIRES USED | | RACE START | |
| | | RACE FINISH | |
| CIRCUIT NAME | | RACING CLASS | |

| TYRE PRESSURE | | START POSITION |
|---|---|---|
| FRONT | REAR | |
| | | LAPS COMPLETED |
| | | |
| | | 1ST ROUND AT |
| | | |

| FASTEST TIME | FASTEST LAP |
|---|---|
| | |
| | |
| | |

## FURTHER OBSERVATIONS

| RACE EVENT | | DATE |
|---|---|---|
| LOCATION | | |
| ORGANISER | | / / |

| RACE NUMBER IN ROW | | CURRENT STANDING | |
|---|---|---|---|
| WEATHER CONDITION | ☼ ☁ ❄ ⛈ ≋ | TIME OF YEAR | 🌷 ☼ 🍂 ❄ |
| TIRES USED | | RACE START / RACE FINISH | 🕒 |
| CIRCUIT NAME | 🏁 | RACING CLASS | |

| TYRE PRESSURE | | START POSITION |
|---|---|---|
| FRONT | REAR | |
| | | LAPS COMPLETED |
| | | |
| | | 1ST ROUND AT |
| | | |

| FASTEST TIME | FASTEST LAP |
|---|---|
| | |
| | |
| | |

FURTHER OBSERVATIONS

| | | | |
|---|---|---|---|
| RACE EVENT | | DATE | |
| LOCATION | | | |
| ORGANISER | | / / | |
| RACE NUMBER IN ROW | | CURRENT STANDING | |
| WEATHER CONDITION | ☼ ☁ ❄ ⚡ ≋ | TIME OF YEAR | 🌷 ☼ 🍂 ❄ |
| TIRES USED | | RACE START / RACE FINISH | 🕒 |
| CIRCUIT NAME | 🏁 | RACING CLASS | |

| TYRE PRESSURE | |
|---|---|
| FRONT | REAR |
| | |
| | |

START POSITION

LAPS COMPLETED

1ST ROUND AT

FASTEST TIME

FASTEST LAP

FURTHER OBSERVATIONS

| | | |
|---|---|---|
| RACE EVENT | | DATE |
| LOCATION | | |
| ORGANISER | | / / |

| | | | |
|---|---|---|---|
| RACE NUMBER IN ROW | | CURRENT STANDING | |
| WEATHER CONDITION | ☼ ☁ ❄ ⛈ ≋ | TIME OF YEAR | 🌷 ☼ 🍂 ❄ |
| TIRES USED | | RACE START / RACE FINISH | 🕒 |
| CIRCUIT NAME | 🏁 | RACING CLASS | |

| TYRE PRESSURE | | START POSITION |
|---|---|---|
| FRONT | REAR | |
| | | LAPS COMPLETED |
| | | |
| | | 1ST ROUND AT |
| | | |

| FASTEST TIME | FASTEST LAP |
|---|---|
| | |
| | |
| | |

FURTHER OBSERVATIONS

| | | |
|---|---|---|
| RACE EVENT | | DATE |
| LOCATION | | |
| ORGANISER | | / / |

| | | | |
|---|---|---|---|
| RACE NUMBER IN ROW | | CURRENT STANDING | |
| WEATHER CONDITION | | TIME OF YEAR | |
| TIRES USED | | RACE START | |
| | | RACE FINISH | |
| CIRCUIT NAME | | RACING CLASS | |

| TYRE PRESSURE | | START POSITION |
|---|---|---|
| FRONT | REAR | |
| | | LAPS COMPLETED |
| | | |
| | | 1ST ROUND AT |
| | | |

| FASTEST TIME | FASTEST LAP |
|---|---|
| | |

FURTHER OBSERVATIONS

| | | |
|---|---|---|
| RACE EVENT | | DATE |
| LOCATION | | |
| ORGANISER | | / / |

| | | | |
|---|---|---|---|
| RACE NUMBER IN ROW | | CURRENT STANDING | |
| WEATHER CONDITION | ☼ ☁ ❄ ⛈ ≋ | TIME OF YEAR | 🌷 ☼ 🍂 ❄ |
| TIRES USED | | RACE START / RACE FINISH | 🕒 |
| CIRCUIT NAME | 🏁 | RACING CLASS | |

| TYRE PRESSURE | | START POSITION |
|---|---|---|
| FRONT | REAR | |
| | | LAPS COMPLETED |
| | | |
| | | 1ST ROUND AT |
| | | |

| FASTEST TIME | FASTEST LAP |
|---|---|
| | |
| | |
| | |

## FURTHER OBSERVATIONS

| | | |
|---|---|---|
| RACE EVENT | | DATE |
| LOCATION | | |
| ORGANISER | | / / |

| | | | |
|---|---|---|---|
| RACE NUMBER IN ROW | | CURRENT STANDING | |
| WEATHER CONDITION | | TIME OF YEAR | |
| TIRES USED | | RACE START | |
| | | RACE FINISH | |
| CIRCUIT NAME | | RACING CLASS | |

| TYRE PRESSURE | | START POSITION |
|---|---|---|
| FRONT | REAR | |
| | | LAPS COMPLETED |
| | | |
| | | 1ST ROUND AT |
| | | |

| FASTEST TIME | FASTEST LAP |
|---|---|
| | |
| | |
| | |

FURTHER OBSERVATIONS

| | | |
|---|---|---|
| RACE EVENT | | DATE |
| LOCATION | | |
| ORGANISER | | / / |

| | | | |
|---|---|---|---|
| RACE NUMBER IN ROW | | CURRENT STANDING | |
| WEATHER CONDITION | | TIME OF YEAR | |
| TIRES USED | | RACE START | |
| | | RACE FINISH | |
| CIRCUIT NAME | | RACING CLASS | |

| TYRE PRESSURE | | START POSITION |
|---|---|---|
| FRONT | REAR | |
| | | LAPS COMPLETED |
| | | |
| | | 1ST ROUND AT |
| | | |

| FASTEST TIME | FASTEST LAP |
|---|---|
| | |
| | |
| | |

| FURTHER OBSERVATIONS |
|---|
| |
| |
| |

| RACE EVENT | | DATE |
|---|---|---|
| LOCATION | | |
| ORGANISER | | / / |

| RACE NUMBER IN ROW | | CURRENT STANDING | |
|---|---|---|---|
| WEATHER CONDITION | | TIME OF YEAR | |
| TIRES USED | | RACE START | |
| | | RACE FINISH | |
| CIRCUIT NAME | | RACING CLASS | |

| TYRE PRESSURE | | START POSITION |
|---|---|---|
| FRONT | REAR | |
| | | LAPS COMPLETED |
| | | |
| | | 1ST ROUND AT |
| | | |

| FASTEST TIME | FASTEST LAP |
|---|---|
| | |
| | |
| | |

FURTHER OBSERVATIONS

| RACE EVENT | | DATE |
|---|---|---|
| LOCATION | | / / |
| ORGANISER | | |

| RACE NUMBER IN ROW | | CURRENT STANDING | |
|---|---|---|---|
| WEATHER CONDITION | | TIME OF YEAR | |
| TIRES USED | | RACE START | |
| | | RACE FINISH | |
| CIRCUIT NAME | | RACING CLASS | |

| TYRE PRESSURE | | START POSITION |
|---|---|---|
| FRONT | REAR | |
| | | LAPS COMPLETED |
| | | |
| | | 1ST ROUND AT |
| | | |

| FASTEST TIME | FASTEST LAP |
|---|---|
| | |
| | |
| | |

FURTHER OBSERVATIONS

| RACE EVENT | | DATE |
|---|---|---|
| LOCATION | | |
| ORGANISER | | / / |

| RACE NUMBER IN ROW | | CURRENT STANDING | |
|---|---|---|---|
| WEATHER CONDITION | | TIME OF YEAR | |
| TIRES USED | | RACE START | |
| | | RACE FINISH | |
| CIRCUIT NAME | | RACING CLASS | |

**TYRE PRESSURE**

| FRONT | REAR |
|---|---|
| | |
| | |

**START POSITION**

**LAPS COMPLETED**

**1ST ROUND AT**

**FASTEST TIME**

**FASTEST LAP**

**FURTHER OBSERVATIONS**

| | | |
|---|---|---|
| RACE EVENT | | DATE |
| LOCATION | | |
| ORGANISER | | / / |

| | | | |
|---|---|---|---|
| RACE NUMBER IN ROW | | CURRENT STANDING | |
| WEATHER CONDITION | | TIME OF YEAR | |
| TIRES USED | | RACE START | |
| | | RACE FINISH | |
| CIRCUIT NAME | | RACING CLASS | |

| TYRE PRESSURE | | START POSITION |
|---|---|---|
| FRONT | REAR | |
| | | LAPS COMPLETED |
| | | |
| | | 1ST ROUND AT |
| | | |

| FASTEST TIME | FASTEST LAP |
|---|---|
| | |
| | |
| | |

| FURTHER OBSERVATIONS |
|---|
| |
| |
| |

| RACE EVENT | | DATE |
|---|---|---|
| LOCATION | | |
| ORGANISER | | / / |

| | | | |
|---|---|---|---|
| RACE NUMBER IN ROW | | CURRENT STANDING | |
| WEATHER CONDITION | | TIME OF YEAR | |
| TIRES USED | | RACE START / RACE FINISH | |
| CIRCUIT NAME | | RACING CLASS | |

**TYRE PRESSURE**

| FRONT | REAR |
|---|---|
| | |
| | |

**START POSITION**

**LAPS COMPLETED**

**1ST ROUND AT**

**FASTEST TIME**

**FASTEST LAP**

**FURTHER OBSERVATIONS**

| | | |
|---|---|---|
| RACE EVENT | | DATE |
| LOCATION | | |
| ORGANISER | | / / |

| | | | |
|---|---|---|---|
| RACE NUMBER IN ROW | | CURRENT STANDING | |
| WEATHER CONDITION | | TIME OF YEAR | |
| TIRES USED | | RACE START | |
| | | RACE FINISH | |
| CIRCUIT NAME | | RACING CLASS | |

| TYRE PRESSURE | |
|---|---|
| FRONT | REAR |
| | |
| | |

START POSITION

LAPS COMPLETED

1ST ROUND AT

| FASTEST TIME | FASTEST LAP |
|---|---|
| | |
| | |
| | |

FURTHER OBSERVATIONS

| RACE EVENT | | DATE |
|---|---|---|
| LOCATION | | |
| ORGANISER | | / / |

| RACE NUMBER IN ROW | | CURRENT STANDING | |
|---|---|---|---|
| WEATHER CONDITION | | TIME OF YEAR | |
| TIRES USED | | RACE START<br>RACE FINISH | |
| CIRCUIT NAME | | RACING CLASS | |

| TYRE PRESSURE | | START POSITION |
|---|---|---|
| FRONT | REAR | |
| | | LAPS COMPLETED |
| | | |
| | | 1ST ROUND AT |
| | | |

| FASTEST TIME | FASTEST LAP |
|---|---|
| | |
| | |
| | |

FURTHER OBSERVATIONS

| | | |
|---|---|---|
| RACE EVENT | | DATE |
| LOCATION | | |
| ORGANISER | | / / |

| | | | |
|---|---|---|---|
| RACE NUMBER IN ROW | | CURRENT STANDING | |
| WEATHER CONDITION | | TIME OF YEAR | |
| TIRES USED | | RACE START | |
| | | RACE FINISH | |
| CIRCUIT NAME | | RACING CLASS | |

| TYRE PRESSURE | | START POSITION |
|---|---|---|
| FRONT | REAR | |
| | | LAPS COMPLETED |
| | | |
| | | 1ST ROUND AT |
| | | |

| FASTEST TIME | FASTEST LAP |
|---|---|
| | |
| | |
| | |

FURTHER OBSERVATIONS

| | | |
|---|---|---|
| RACE EVENT | | DATE |
| LOCATION | | |
| ORGANISER | | / / |

| | | | |
|---|---|---|---|
| RACE NUMBER IN ROW | | CURRENT STANDING | |
| WEATHER CONDITION | | TIME OF YEAR | |
| TIRES USED | | RACE START | |
| | | RACE FINISH | |
| CIRCUIT NAME | | RACING CLASS | |

| TYRE PRESSURE | |
|---|---|
| FRONT | REAR |
| | |
| | |

START POSITION

LAPS COMPLETED

1ST ROUND AT

| FASTEST TIME | FASTEST LAP |
|---|---|
| | |

FURTHER OBSERVATIONS

| | | |
|---|---|---|
| RACE EVENT | | DATE |
| LOCATION | | / / |
| ORGANISER | | |

| | | | |
|---|---|---|---|
| RACE NUMBER IN ROW | | CURRENT STANDING | |
| WEATHER CONDITION | | TIME OF YEAR | |
| TIRES USED | | RACE START | |
| | | RACE FINISH | |
| CIRCUIT NAME | | RACING CLASS | |

| TYRE PRESSURE | |
|---|---|
| FRONT | REAR |
| | |
| | |

START POSITION

LAPS COMPLETED

1ST ROUND AT

| FASTEST TIME | FASTEST LAP |
|---|---|
| | |

FURTHER OBSERVATIONS

| RACE EVENT | | DATE |
|---|---|---|
| LOCATION | | |
| ORGANISER | | / / |

| RACE NUMBER IN ROW | | CURRENT STANDING | |
|---|---|---|---|
| WEATHER CONDITION | ☼ ☁ ❄ ⛈ ≋ | TIME OF YEAR | 🌷 ☼ 🍂 ❄ |
| TIRES USED | | RACE START / RACE FINISH | 🕒 |
| CIRCUIT NAME | 🏁 | RACING CLASS | |

| TYRE PRESSURE | |
|---|---|
| FRONT | REAR |
| | |
| | |

START POSITION

LAPS COMPLETED

1ST ROUND AT

| FASTEST TIME | FASTEST LAP |
|---|---|
| | |
| | |
| | |

FURTHER OBSERVATIONS

| | | |
|---|---|---|
| RACE EVENT | | DATE |
| LOCATION | | |
| ORGANISER | | / / |

| | | | |
|---|---|---|---|
| RACE NUMBER IN ROW | | CURRENT STANDING | |
| WEATHER CONDITION | | TIME OF YEAR | |
| TIRES USED | | RACE START | |
| | | RACE FINISH | |
| CIRCUIT NAME | | RACING CLASS | |

| TYRE PRESSURE | |
|---|---|
| FRONT | REAR |
| | |
| | |

START POSITION

LAPS COMPLETED

1ST ROUND AT

| FASTEST TIME | FASTEST LAP |
|---|---|
| | |

FURTHER OBSERVATIONS

| | | |
|---|---|---|
| RACE EVENT | | DATE |
| LOCATION | | |
| ORGANISER | | / / |

| | | | |
|---|---|---|---|
| RACE NUMBER IN ROW | | CURRENT STANDING | |
| WEATHER CONDITION | | TIME OF YEAR | |
| TIRES USED | | RACE START<br>RACE FINISH | |
| CIRCUIT NAME | | RACING CLASS | |

| TYRE PRESSURE | | START POSITION |
|---|---|---|
| FRONT | REAR | |
| | | LAPS COMPLETED |
| | | |
| | | 1ST ROUND AT |
| | | |

| FASTEST TIME | FASTEST LAP |
|---|---|
| | |
| | |
| | |

| FURTHER OBSERVATIONS |
|---|
| |
| |
| |

| | | |
|---|---|---|
| RACE EVENT | | DATE |
| LOCATION | | |
| ORGANISER | | / / |

| | | | |
|---|---|---|---|
| RACE NUMBER IN ROW | | CURRENT STANDING | |
| WEATHER CONDITION | | TIME OF YEAR | |
| TIRES USED | | RACE START | |
| | | RACE FINISH | |
| CIRCUIT NAME | | RACING CLASS | |

| TYRE PRESSURE | | START POSITION |
|---|---|---|
| FRONT | REAR | |
| | | LAPS COMPLETED |
| | | |
| | | 1ST ROUND AT |
| | | |

| FASTEST TIME | FASTEST LAP |
|---|---|
| | |
| | |
| | |

FURTHER OBSERVATIONS

| RACE EVENT | | DATE |
|---|---|---|
| LOCATION | | |
| ORGANISER | | / / |

| RACE NUMBER IN ROW | | CURRENT STANDING | |
|---|---|---|---|
| WEATHER CONDITION | | TIME OF YEAR | |
| TIRES USED | | RACE START | |
| | | RACE FINISH | |
| CIRCUIT NAME | | RACING CLASS | |

| TYRE PRESSURE | | START POSITION |
|---|---|---|
| FRONT | REAR | |
| | | LAPS COMPLETED |
| | | |
| | | 1ST ROUND AT |
| | | |

| FASTEST TIME | FASTEST LAP |
|---|---|
| | |
| | |
| | |

| FURTHER OBSERVATIONS |
|---|
| |
| |
| |

| | | |
|---|---|---|
| RACE EVENT | | DATE |
| LOCATION | | |
| ORGANISER | | / / |

| | | | |
|---|---|---|---|
| RACE NUMBER IN ROW | | CURRENT STANDING | |
| WEATHER CONDITION | | TIME OF YEAR | |
| TIRES USED | | RACE START | |
| | | RACE FINISH | |
| CIRCUIT NAME | | RACING CLASS | |

| TYRE PRESSURE | | START POSITION |
|---|---|---|
| FRONT | REAR | |
| | | LAPS COMPLETED |
| | | |
| | | 1ST ROUND AT |
| | | |

| FASTEST TIME | FASTEST LAP |
|---|---|
| | |
| | |
| | |

FURTHER OBSERVATIONS

| | | |
|---|---|---|
| RACE EVENT | | DATE |
| LOCATION | | / / |
| ORGANISER | | |

| | | | |
|---|---|---|---|
| RACE NUMBER IN ROW | | CURRENT STANDING | |
| WEATHER CONDITION | | TIME OF YEAR | |
| TIRES USED | | RACE START | |
| | | RACE FINISH | |
| CIRCUIT NAME | | RACING CLASS | |

| TYRE PRESSURE | | START POSITION |
|---|---|---|
| FRONT | REAR | |
| | | LAPS COMPLETED |
| | | |
| | | 1ST ROUND AT |
| | | |

| FASTEST TIME | FASTEST LAP |
|---|---|
| | |
| | |
| | |

FURTHER OBSERVATIONS

| | | |
|---|---|---|
| RACE EVENT | | DATE |
| LOCATION | | / / |
| ORGANISER | | |

| | | | |
|---|---|---|---|
| RACE NUMBER IN ROW | | CURRENT STANDING | |
| WEATHER CONDITION | ☼ ☁ ❄ ⚡ ≈ | TIME OF YEAR | ❀ ☼ ❦ ❄ |
| TIRES USED | | RACE START<br>RACE FINISH | 🕒 |
| CIRCUIT NAME | 🏁 | RACING CLASS | |

| TYRE PRESSURE | | START POSITION |
|---|---|---|
| FRONT | REAR | |
| | | LAPS COMPLETED |
| | | |
| | | 1ST ROUND AT |
| | | |

| FASTEST TIME | FASTEST LAP |
|---|---|
| | |
| | |
| | |

## FURTHER OBSERVATIONS

| | | |
|---|---|---|
| RACE EVENT | | DATE |
| LOCATION | | |
| ORGANISER | | / / |

| | | | |
|---|---|---|---|
| RACE NUMBER IN ROW | | CURRENT STANDING | |
| WEATHER CONDITION | ☼ ☁ ❄ ⛈ ≋ | TIME OF YEAR | 🌷 ☼ 🍂 ❄ |
| TIRES USED | | RACE START | |
| | | RACE FINISH | |
| CIRCUIT NAME | | RACING CLASS | |

| TYRE PRESSURE | |
|---|---|
| FRONT | REAR |
| | |
| | |

START POSITION

LAPS COMPLETED

1ST ROUND AT

FASTEST TIME

FASTEST LAP

FURTHER OBSERVATIONS

| | | |
|---|---|---|
| RACE EVENT | | DATE |
| LOCATION | | |
| ORGANISER | | / / |

| | | | |
|---|---|---|---|
| RACE NUMBER IN ROW | | CURRENT STANDING | |
| WEATHER CONDITION | | TIME OF YEAR | |
| TIRES USED | | RACE START | |
| | | RACE FINISH | |
| CIRCUIT NAME | | RACING CLASS | |

## TYRE PRESSURE

| FRONT | REAR |
|---|---|
| | |
| | |

## START POSITION

## LAPS COMPLETED

## 1ST ROUND AT

## FASTEST TIME

## FASTEST LAP

## FURTHER OBSERVATIONS

| RACE EVENT | | DATE |
|---|---|---|
| LOCATION | | |
| ORGANISER | | / / |

| RACE NUMBER IN ROW | | CURRENT STANDING | |
|---|---|---|---|
| WEATHER CONDITION | ☼ ☁ ❄ ⛈ ≋ | TIME OF YEAR | 🌷 ☼ 🍂 ❄ |
| TIRES USED | | RACE START / RACE FINISH | 🕒 |
| CIRCUIT NAME | 🏁 | RACING CLASS | |

| TYRE PRESSURE | |
|---|---|
| FRONT | REAR |
| | |
| | |

START POSITION

LAPS COMPLETED

1ST ROUND AT

| FASTEST TIME | FASTEST LAP |
|---|---|
| | |
| | |
| | |

FURTHER OBSERVATIONS

| | | |
|---|---|---|
| RACE EVENT | | DATE |
| LOCATION | | |
| ORGANISER | | / / |

| | | | |
|---|---|---|---|
| RACE NUMBER IN ROW | | CURRENT STANDING | |
| WEATHER CONDITION | | TIME OF YEAR | |
| TIRES USED | | RACE START | |
| | | RACE FINISH | |
| CIRCUIT NAME | | RACING CLASS | |

| TYRE PRESSURE | |
|---|---|
| FRONT | REAR |
| | |
| | |

START POSITION

LAPS COMPLETED

1ST ROUND AT

FASTEST TIME

FASTEST LAP

FURTHER OBSERVATIONS

| | | |
|---|---|---|
| RACE EVENT | | DATE |
| LOCATION | | / / |
| ORGANISER | | |

| | | | |
|---|---|---|---|
| RACE NUMBER IN ROW | | CURRENT STANDING | |
| WEATHER CONDITION | | TIME OF YEAR | |
| TIRES USED | | RACE START | |
| | | RACE FINISH | |
| CIRCUIT NAME | | RACING CLASS | |

| TYRE PRESSURE | | START POSITION |
|---|---|---|
| FRONT | REAR | |
| | | LAPS COMPLETED |
| | | |
| | | 1ST ROUND AT |
| | | |

| FASTEST TIME | FASTEST LAP |
|---|---|
| | |
| | |
| | |

FURTHER OBSERVATIONS

| RACE EVENT | | DATE |
|---|---|---|
| LOCATION | | |
| ORGANISER | | / / |

| RACE NUMBER IN ROW | | CURRENT STANDING | |
|---|---|---|---|
| WEATHER CONDITION | ☼ ☁ ❄ ⛈ ≋ | TIME OF YEAR | 🌷 ☼ 🍂 ❄ |
| TIRES USED | | RACE START / RACE FINISH | 🕒 |
| CIRCUIT NAME | 🏁 | RACING CLASS | |

| TYRE PRESSURE | | START POSITION |
|---|---|---|
| FRONT | REAR | |
| | | LAPS COMPLETED |
| | | |
| | | 1ST ROUND AT |
| | | |

| FASTEST TIME | FASTEST LAP |
|---|---|
| | |
| | |
| | |

| FURTHER OBSERVATIONS |
|---|
| |
| |
| |

| | | |
|---|---|---|
| RACE EVENT | | DATE |
| LOCATION | | |
| ORGANISER | | / / |

| | | | |
|---|---|---|---|
| RACE NUMBER IN ROW | | CURRENT STANDING | |
| WEATHER CONDITION | ☼ ☁ ❄ ⛈ ≋ | TIME OF YEAR | ❀ ☼ ❦ ❄ |
| TIRES USED | | RACE START | |
| | | RACE FINISH | |
| CIRCUIT NAME | | RACING CLASS | |

**TYRE PRESSURE**

| FRONT | REAR |
|---|---|
| | |
| | |

**START POSITION**

**LAPS COMPLETED**

**1ST ROUND AT**

**FASTEST TIME**

**FASTEST LAP**

**FURTHER OBSERVATIONS**

| | | |
|---|---|---|
| RACE EVENT | | DATE |
| LOCATION | | / / |
| ORGANISER | | |

| | | | |
|---|---|---|---|
| RACE NUMBER IN ROW | | CURRENT STANDING | |
| WEATHER CONDITION | | TIME OF YEAR | |
| TIRES USED | | RACE START | |
| | | RACE FINISH | |
| CIRCUIT NAME | | RACING CLASS | |

| TYRE PRESSURE | |
|---|---|
| FRONT | REAR |
| | |
| | |

START POSITION

LAPS COMPLETED

1ST ROUND AT

| FASTEST TIME | FASTEST LAP |
|---|---|
| | |

FURTHER OBSERVATIONS

| RACE EVENT | | DATE |
|---|---|---|
| LOCATION | | |
| ORGANISER | | / / |

| RACE NUMBER IN ROW | | CURRENT STANDING | |
|---|---|---|---|
| WEATHER CONDITION | | TIME OF YEAR | |
| TIRES USED | | RACE START | |
| | | RACE FINISH | |
| CIRCUIT NAME | | RACING CLASS | |

## TYRE PRESSURE

| FRONT | REAR |
|---|---|
| | |
| | |

## START POSITION

## LAPS COMPLETED

## 1ST ROUND AT

## FASTEST TIME

## FASTEST LAP

## FURTHER OBSERVATIONS

| RACE EVENT | | DATE |
|---|---|---|
| LOCATION | | |
| ORGANISER | | / / |

| RACE NUMBER IN ROW | | CURRENT STANDING | |
|---|---|---|---|
| WEATHER CONDITION | | TIME OF YEAR | |
| TIRES USED | | RACE START / RACE FINISH | |
| CIRCUIT NAME | | RACING CLASS | |

| TYRE PRESSURE | |
|---|---|
| FRONT | REAR |
| | |
| | |

START POSITION

LAPS COMPLETED

1ST ROUND AT

| FASTEST TIME | FASTEST LAP |
|---|---|
| | |

FURTHER OBSERVATIONS

| | | |
|---|---|---|
| RACE EVENT | | DATE |
| LOCATION | | |
| ORGANISER | | / / |

| | | | |
|---|---|---|---|
| RACE NUMBER IN ROW | | CURRENT STANDING | |
| WEATHER CONDITION | ☼ ☁ ❄ ⚡ ≋ | TIME OF YEAR | 🌷 ☼ 🍂 ❄ |
| TIRES USED | | RACE START / RACE FINISH | 🕒 |
| CIRCUIT NAME | 🏁 | RACING CLASS | |

| TYRE PRESSURE | |
|---|---|
| FRONT | REAR |
| | |
| | |

START POSITION

LAPS COMPLETED

1ST ROUND AT

| FASTEST TIME | FASTEST LAP |
|---|---|
| | |
| | |
| | |

FURTHER OBSERVATIONS

| | | |
|---|---|---|
| RACE EVENT | | DATE |
| LOCATION | | |
| ORGANISER | | / / |

| | | | |
|---|---|---|---|
| RACE NUMBER IN ROW | | CURRENT STANDING | |
| WEATHER CONDITION | | TIME OF YEAR | |
| TIRES USED | | RACE START | |
| | | RACE FINISH | |
| CIRCUIT NAME | | RACING CLASS | |

| TYRE PRESSURE | |
|---|---|
| FRONT | REAR |
| | |
| | |

START POSITION

LAPS COMPLETED

1ST ROUND AT

| FASTEST TIME | FASTEST LAP |
|---|---|
| | |
| | |
| | |

FURTHER OBSERVATIONS

| | | |
|---|---|---|
| RACE EVENT | | DATE |
| LOCATION | | |
| ORGANISER | | / / |

| | | | |
|---|---|---|---|
| RACE NUMBER IN ROW | | CURRENT STANDING | |
| WEATHER CONDITION | | TIME OF YEAR | |
| TIRES USED | | RACE START | |
| | | RACE FINISH | |
| CIRCUIT NAME | | RACING CLASS | |

| TYRE PRESSURE | |
|---|---|
| FRONT | REAR |
| | |
| | |

START POSITION

LAPS COMPLETED

1ST ROUND AT

| FASTEST TIME | FASTEST LAP |
|---|---|
| | |

FURTHER OBSERVATIONS

| | | | |
|---|---|---|---|
| RACE EVENT | | DATE | |
| LOCATION | | / / | |
| ORGANISER | | | |
| RACE NUMBER IN ROW | | CURRENT STANDING | |
| WEATHER CONDITION | | TIME OF YEAR | |
| TIRES USED | | RACE START | |
| | | RACE FINISH | |
| CIRCUIT NAME | | RACING CLASS | |

## TYRE PRESSURE

| FRONT | REAR |
|---|---|
| | |
| | |

## START POSITION

## LAPS COMPLETED

## 1ST ROUND AT

## FASTEST TIME

## FASTEST LAP

## FURTHER OBSERVATIONS

| RACE EVENT | | DATE |
|---|---|---|
| LOCATION | | |
| ORGANISER | | / / |

| RACE NUMBER IN ROW | | CURRENT STANDING | |
|---|---|---|---|
| WEATHER CONDITION | | TIME OF YEAR | |
| TIRES USED | | RACE START | |
| | | RACE FINISH | |
| CIRCUIT NAME | | RACING CLASS | |

| TYRE PRESSURE | |
|---|---|
| FRONT | REAR |
| | |
| | |

START POSITION

LAPS COMPLETED

1ST ROUND AT

| FASTEST TIME | FASTEST LAP |
|---|---|
| | |
| | |
| | |

FURTHER OBSERVATIONS

| RACE EVENT | | DATE |
|---|---|---|
| LOCATION | | / / |
| ORGANISER | | |

| RACE NUMBER IN ROW | | CURRENT STANDING | |
|---|---|---|---|
| WEATHER CONDITION | | TIME OF YEAR | |
| TIRES USED | | RACE START / RACE FINISH | |
| CIRCUIT NAME | | RACING CLASS | |

| TYRE PRESSURE | | START POSITION |
|---|---|---|
| FRONT | REAR | |
| | | LAPS COMPLETED |
| | | |
| | | 1ST ROUND AT |
| | | |

| FASTEST TIME | FASTEST LAP |
|---|---|
| | |
| | |
| | |

| FURTHER OBSERVATIONS |
|---|
| |
| |
| |

| RACE EVENT | | DATE |
|---|---|---|
| LOCATION | | / / |
| ORGANISER | | |

| RACE NUMBER IN ROW | | CURRENT STANDING | |
|---|---|---|---|
| WEATHER CONDITION | | TIME OF YEAR | |
| TIRES USED | | RACE START / RACE FINISH | |
| CIRCUIT NAME | | RACING CLASS | |

| TYRE PRESSURE | |
|---|---|
| FRONT | REAR |
| | |
| | |

START POSITION

LAPS COMPLETED

1ST ROUND AT

| FASTEST TIME | FASTEST LAP |
|---|---|
| | |

FURTHER OBSERVATIONS

| RACE EVENT | | DATE |
|---|---|---|
| LOCATION | | / / |
| ORGANISER | | |

| RACE NUMBER IN ROW | | CURRENT STANDING | |
|---|---|---|---|
| WEATHER CONDITION | ☼ ☁ ❄ ⛈ ≋ | TIME OF YEAR | 🌷 ☼ 🍂 ❄ |
| TIRES USED | | RACE START / RACE FINISH | 🕒 |
| CIRCUIT NAME | 🏁 | RACING CLASS | |

**TYRE PRESSURE**

| FRONT | REAR |
|---|---|
| | |
| | |

**START POSITION**

**LAPS COMPLETED**

**1ST ROUND AT**

**FASTEST TIME**

**FASTEST LAP**

**FURTHER OBSERVATIONS**

| | | |
|---|---|---|
| RACE EVENT | | DATE |
| LOCATION | | |
| ORGANISER | | / / |

| | | | |
|---|---|---|---|
| RACE NUMBER IN ROW | | CURRENT STANDING | |
| WEATHER CONDITION | | TIME OF YEAR | |
| TIRES USED | | RACE START | |
| | | RACE FINISH | |
| CIRCUIT NAME | | RACING CLASS | |

| TYRE PRESSURE | | START POSITION |
|---|---|---|
| FRONT | REAR | |
| | | LAPS COMPLETED |
| | | |
| | | 1ST ROUND AT |
| | | |

| FASTEST TIME | FASTEST LAP |
|---|---|
| | |
| | |
| | |

FURTHER OBSERVATIONS

| | | |
|---|---|---|
| RACE EVENT | | DATE |
| LOCATION | | |
| ORGANISER | | / / |

| | | | |
|---|---|---|---|
| RACE NUMBER IN ROW | | CURRENT STANDING | |
| WEATHER CONDITION | | TIME OF YEAR | |
| TIRES USED | | RACE START | |
| | | RACE FINISH | |
| CIRCUIT NAME | | RACING CLASS | |

| TYRE PRESSURE | |
|---|---|
| FRONT | REAR |
| | |
| | |

START POSITION

LAPS COMPLETED

1ST ROUND AT

| FASTEST TIME | FASTEST LAP |
|---|---|
| | |
| | |
| | |

FURTHER OBSERVATIONS

| | | DATE |
|---|---|---|
| RACE EVENT | | / / |
| LOCATION | | |
| ORGANISER | | |

| | | | |
|---|---|---|---|
| RACE NUMBER IN ROW | | CURRENT STANDING | |
| WEATHER CONDITION | | TIME OF YEAR | |
| TIRES USED | | RACE START | |
| | | RACE FINISH | |
| CIRCUIT NAME | | RACING CLASS | |

| TYRE PRESSURE | | START POSITION |
|---|---|---|
| FRONT | REAR | |
| | | LAPS COMPLETED |
| | | |
| | | 1ST ROUND AT |
| | | |

| FASTEST TIME | FASTEST LAP |
|---|---|
| | |
| | |
| | |

FURTHER OBSERVATIONS

| RACE EVENT | | DATE |
|---|---|---|
| LOCATION | | |
| ORGANISER | | / / |

| RACE NUMBER IN ROW | | CURRENT STANDING | |
|---|---|---|---|
| WEATHER CONDITION | ☼ ☁ ❄ ⛈ ≋ | TIME OF YEAR | 🌷 ☼ 🍂 ❄ |
| TIRES USED | | RACE START / RACE FINISH | 🕒 |
| CIRCUIT NAME | 🏁 | RACING CLASS | |

## TYRE PRESSURE

| FRONT | REAR |
|---|---|
| | |
| | |

## START POSITION

## LAPS COMPLETED

## 1ST ROUND AT

## FASTEST TIME

## FASTEST LAP

## FURTHER OBSERVATIONS

| RACE EVENT | | DATE |
|---|---|---|
| LOCATION | | |
| ORGANISER | | / / |

| RACE NUMBER IN ROW | | CURRENT STANDING | |
|---|---|---|---|
| WEATHER CONDITION | | TIME OF YEAR | |
| TIRES USED | | RACE START<br>RACE FINISH | |
| CIRCUIT NAME | | RACING CLASS | |

| TYRE PRESSURE | |
|---|---|
| FRONT | REAR |
| | |
| | |

START POSITION

LAPS COMPLETED

1ST ROUND AT

| FASTEST TIME | FASTEST LAP |
|---|---|
| | |

FURTHER OBSERVATIONS

| | | |
|---|---|---|
| RACE EVENT | | DATE |
| LOCATION | | / / |
| ORGANISER | | |

| | | | |
|---|---|---|---|
| RACE NUMBER IN ROW | | CURRENT STANDING | |
| WEATHER CONDITION | | TIME OF YEAR | |
| TIRES USED | | RACE START | |
| | | RACE FINISH | |
| CIRCUIT NAME | | RACING CLASS | |

| TYRE PRESSURE | | START POSITION |
|---|---|---|
| FRONT | REAR | |
| | | LAPS COMPLETED |
| | | |
| | | 1ST ROUND AT |
| | | |

| FASTEST TIME | FASTEST LAP |
|---|---|
| | |
| | |
| | |

FURTHER OBSERVATIONS

| RACE EVENT | | DATE |
|---|---|---|
| LOCATION | | |
| ORGANISER | | / / |

| RACE NUMBER IN ROW | | CURRENT STANDING | |
|---|---|---|---|
| WEATHER CONDITION | | TIME OF YEAR | |
| TIRES USED | | RACE START | |
| | | RACE FINISH | |
| CIRCUIT NAME | | RACING CLASS | |

| TYRE PRESSURE | |
|---|---|
| FRONT | REAR |
| | |
| | |

START POSITION

LAPS COMPLETED

1ST ROUND AT

| FASTEST TIME | FASTEST LAP |
|---|---|
| | |
| | |
| | |

FURTHER OBSERVATIONS

| RACE EVENT | | DATE |
|---|---|---|
| LOCATION | | / / |
| ORGANISER | | |

| | | | |
|---|---|---|---|
| RACE NUMBER IN ROW | | CURRENT STANDING | |
| WEATHER CONDITION | ☼ ☁ ❄ ⚡ ≋ | TIME OF YEAR | 🌷 ☼ 🍂 ❄ |
| TIRES USED | | RACE START / RACE FINISH | 🕒 |
| CIRCUIT NAME | 🏁 | RACING CLASS | |

| TYRE PRESSURE | |
|---|---|
| FRONT | REAR |
| | |
| | |

START POSITION

LAPS COMPLETED

1ST ROUND AT

| FASTEST TIME | FASTEST LAP |
|---|---|
| | |
| | |
| | |

FURTHER OBSERVATIONS

| RACE EVENT | | DATE |
|---|---|---|
| LOCATION | | |
| ORGANISER | | / / |

| | | | |
|---|---|---|---|
| RACE NUMBER IN ROW | | CURRENT STANDING | |
| WEATHER CONDITION | | TIME OF YEAR | |
| TIRES USED | | RACE START | |
| | | RACE FINISH | |
| CIRCUIT NAME | | RACING CLASS | |

| TYRE PRESSURE | | START POSITION |
|---|---|---|
| FRONT | REAR | |
| | | LAPS COMPLETED |
| | | |
| | | 1ST ROUND AT |
| | | |

| FASTEST TIME | FASTEST LAP |
|---|---|
| | |
| | |
| | |

| FURTHER OBSERVATIONS |
|---|
| |
| |
| |

| RACE EVENT | | DATE |
|---|---|---|
| LOCATION | | |
| ORGANISER | | / / |

| RACE NUMBER IN ROW | | CURRENT STANDING | |
|---|---|---|---|
| WEATHER CONDITION | | TIME OF YEAR | |
| TIRES USED | | RACE START | |
| | | RACE FINISH | |
| CIRCUIT NAME | | RACING CLASS | |

| TYRE PRESSURE | | START POSITION |
|---|---|---|
| FRONT | REAR | |
| | | LAPS COMPLETED |
| | | |
| | | 1ST ROUND AT |
| | | |

| FASTEST TIME | FASTEST LAP |
|---|---|
| | |
| | |
| | |

FURTHER OBSERVATIONS

| | | |
|---|---|---|
| RACE EVENT | | DATE |
| LOCATION | | |
| ORGANISER | | / / |

| | | | |
|---|---|---|---|
| RACE NUMBER IN ROW | | CURRENT STANDING | |
| WEATHER CONDITION | | TIME OF YEAR | |
| TIRES USED | | RACE START | |
| | | RACE FINISH | |
| CIRCUIT NAME | | RACING CLASS | |

## TYRE PRESSURE

| FRONT | REAR |
|---|---|
| | |
| | |

## START POSITION

## LAPS COMPLETED

## 1ST ROUND AT

## FASTEST TIME

## FASTEST LAP

## FURTHER OBSERVATIONS

| | | |
|---|---|---|
| RACE EVENT | | DATE |
| LOCATION | | |
| ORGANISER | | / / |

| | | | |
|---|---|---|---|
| RACE NUMBER IN ROW | | CURRENT STANDING | |
| WEATHER CONDITION | | TIME OF YEAR | |
| TIRES USED | | RACE START<br>RACE FINISH | |
| CIRCUIT NAME | | RACING CLASS | |

| TYRE PRESSURE | | START POSITION |
|---|---|---|
| FRONT | REAR | |
| | | LAPS COMPLETED |
| | | |
| | | 1ST ROUND AT |
| | | |

| FASTEST TIME | FASTEST LAP |
|---|---|
| | |
| | |
| | |

FURTHER OBSERVATIONS

| | | |
|---|---|---|
| RACE EVENT | | DATE |
| LOCATION | | |
| ORGANISER | | / / |

| | | | |
|---|---|---|---|
| RACE NUMBER IN ROW | | CURRENT STANDING | |
| WEATHER CONDITION | ☀ 🌧 ❄ ⛈ 💨 | TIME OF YEAR | 🌷 ☀ 🍂 ❄ |
| TIRES USED | | RACE START<br>RACE FINISH | 🕒 |
| CIRCUIT NAME | 🏁 | RACING CLASS | |

| TYRE PRESSURE | | START POSITION |
|---|---|---|
| FRONT | REAR | |
| | | LAPS COMPLETED |
| | | |
| | | 1ST ROUND AT |
| | | |

| FASTEST TIME | FASTEST LAP |
|---|---|
| | |
| | |
| | |

| FURTHER OBSERVATIONS |
|---|
| |
| |
| |

| | | |
|---|---|---|
| RACE EVENT | | DATE |
| LOCATION | | |
| ORGANISER | | / / |

| | | | |
|---|---|---|---|
| RACE NUMBER IN ROW | | CURRENT STANDING | |
| WEATHER CONDITION | | TIME OF YEAR | |
| TIRES USED | | RACE START<br>RACE FINISH | |
| CIRCUIT NAME | | RACING CLASS | |

| TYRE PRESSURE | | START POSITION |
|---|---|---|
| FRONT | REAR | |
| | | LAPS COMPLETED |
| | | |
| | | 1ST ROUND AT |
| | | |

| FASTEST TIME | FASTEST LAP |
|---|---|
| | |
| | |
| | |

| FURTHER OBSERVATIONS |
|---|
| |
| |
| |

| | | |
|---|---|---|
| RACE EVENT | | DATE |
| LOCATION | | |
| ORGANISER | | / / |

| | | | |
|---|---|---|---|
| RACE NUMBER IN ROW | | CURRENT STANDING | |
| WEATHER CONDITION | | TIME OF YEAR | |
| TIRES USED | | RACE START | |
| | | RACE FINISH | |
| CIRCUIT NAME | | RACING CLASS | |

| TYRE PRESSURE | | START POSITION |
|---|---|---|
| FRONT | REAR | |
| | | LAPS COMPLETED |
| | | |
| | | 1ST ROUND AT |
| | | |

| FASTEST TIME | FASTEST LAP |
|---|---|
| | |
| | |
| | |

| FURTHER OBSERVATIONS |
|---|
| |
| |
| |

| RACE EVENT | | DATE |
|---|---|---|
| LOCATION | | |
| ORGANISER | | / / |

| RACE NUMBER IN ROW | | CURRENT STANDING | |
|---|---|---|---|
| WEATHER CONDITION | | TIME OF YEAR | |
| TIRES USED | | RACE START | |
| | | RACE FINISH | |
| CIRCUIT NAME | | RACING CLASS | |

| TYRE PRESSURE | |
|---|---|
| FRONT | REAR |
| | |
| | |

START POSITION

LAPS COMPLETED

1ST ROUND AT

| FASTEST TIME | FASTEST LAP |
|---|---|
| | |

FURTHER OBSERVATIONS

| RACE EVENT | | DATE |
|---|---|---|
| LOCATION | | |
| ORGANISER | | / / |

| RACE NUMBER IN ROW | | CURRENT STANDING | |
|---|---|---|---|
| WEATHER CONDITION | | TIME OF YEAR | |
| TIRES USED | | RACE START | |
| | | RACE FINISH | |
| CIRCUIT NAME | | RACING CLASS | |

| TYRE PRESSURE | | START POSITION |
|---|---|---|
| FRONT | REAR | |
| | | LAPS COMPLETED |
| | | |
| | | 1ST ROUND AT |
| | | |

| FASTEST TIME | FASTEST LAP |
|---|---|
| | |
| | |
| | |

| FURTHER OBSERVATIONS |
|---|
| |
| |
| |

| | | |
|---|---|---|
| RACE EVENT | | DATE |
| LOCATION | | |
| ORGANISER | | / / |

| | | | |
|---|---|---|---|
| RACE NUMBER IN ROW | | CURRENT STANDING | |
| WEATHER CONDITION | | TIME OF YEAR | |
| TIRES USED | | RACE START | |
| | | RACE FINISH | |
| CIRCUIT NAME | | RACING CLASS | |

| TYRE PRESSURE | | START POSITION |
|---|---|---|
| FRONT | REAR | |
| | | LAPS COMPLETED |
| | | |
| | | 1ST ROUND AT |
| | | |

| FASTEST TIME | FASTEST LAP |
|---|---|
| | |
| | |
| | |

FURTHER OBSERVATIONS

| | | |
|---|---|---|
| RACE EVENT | | DATE |
| LOCATION | | |
| ORGANISER | | / / |

| | | | |
|---|---|---|---|
| RACE NUMBER IN ROW | | CURRENT STANDING | |
| WEATHER CONDITION | ☼ ☁ ❄ ⛈ ≋ | TIME OF YEAR | ✿ ☼ ❦ ❄ |
| TIRES USED | | RACE START / RACE FINISH | ◷ |
| CIRCUIT NAME | ⚑ | RACING CLASS | |

| TYRE PRESSURE | | START POSITION |
|---|---|---|
| FRONT | REAR | |
| | | LAPS COMPLETED |
| | | |
| | | 1ST ROUND AT |
| | | |

| FASTEST TIME | FASTEST LAP |
|---|---|
| | |
| | |
| | |

FURTHER OBSERVATIONS

| RACE EVENT | | DATE |
|---|---|---|
| LOCATION | | / / |
| ORGANISER | | |

| RACE NUMBER IN ROW | | CURRENT STANDING | |
|---|---|---|---|
| WEATHER CONDITION | | TIME OF YEAR | |
| TIRES USED | | RACE START / RACE FINISH | |
| CIRCUIT NAME | | RACING CLASS | |

## TYRE PRESSURE

| FRONT | REAR |
|---|---|
| | |
| | |

## START POSITION

## LAPS COMPLETED

## 1ST ROUND AT

## FASTEST TIME

## FASTEST LAP

## FURTHER OBSERVATIONS

| | | |
|---|---|---|
| RACE EVENT | | DATE |
| LOCATION | | |
| ORGANISER | | / / |

| | | | |
|---|---|---|---|
| RACE NUMBER IN ROW | | CURRENT STANDING | |
| WEATHER CONDITION | | TIME OF YEAR | |
| TIRES USED | | RACE START | |
| | | RACE FINISH | |
| CIRCUIT NAME | | RACING CLASS | |

| TYRE PRESSURE | | START POSITION |
|---|---|---|
| FRONT | REAR | |
| | | LAPS COMPLETED |
| | | |
| | | 1ST ROUND AT |
| | | |

| FASTEST TIME | FASTEST LAP |
|---|---|
| | |
| | |
| | |

**FURTHER OBSERVATIONS**

| | | |
|---|---|---|
| RACE EVENT | | DATE |
| LOCATION | | |
| ORGANISER | | / / |

| | | | |
|---|---|---|---|
| RACE NUMBER IN ROW | | CURRENT STANDING | |
| WEATHER CONDITION | | TIME OF YEAR | |
| TIRES USED | | RACE START | |
| | | RACE FINISH | |
| CIRCUIT NAME | | RACING CLASS | |

| TYRE PRESSURE | |
|---|---|
| FRONT | REAR |
| | |
| | |

START POSITION

LAPS COMPLETED

1ST ROUND AT

| FASTEST TIME | FASTEST LAP |
|---|---|
| | |
| | |
| | |

FURTHER OBSERVATIONS

| RACE EVENT | | DATE |
|---|---|---|
| LOCATION | | / / |
| ORGANISER | | |

| RACE NUMBER IN ROW | | CURRENT STANDING | |
|---|---|---|---|
| WEATHER CONDITION | | TIME OF YEAR | |
| TIRES USED | | RACE START | |
| | | RACE FINISH | |
| CIRCUIT NAME | | RACING CLASS | |

TYRE PRESSURE

| FRONT | REAR |
|---|---|
| | |
| | |

START POSITION

LAPS COMPLETED

1ST ROUND AT

FASTEST TIME

FASTEST LAP

FURTHER OBSERVATIONS

| | | |
|---|---|---|
| RACE EVENT | | DATE |
| LOCATION | | |
| ORGANISER | | / / |

| | | | |
|---|---|---|---|
| RACE NUMBER IN ROW | | CURRENT STANDING | |
| WEATHER CONDITION | ☼ ☁ ❄ ⛈ ≋ | TIME OF YEAR | 🌷 ☼ 🍂 ❄ |
| TIRES USED | | RACE START | 🕒 |
| | | RACE FINISH | |
| CIRCUIT NAME | 🏁 | RACING CLASS | |

| TYRE PRESSURE | | START POSITION |
|---|---|---|
| FRONT | REAR | |
| | | LAPS COMPLETED |
| | | |
| | | 1ST ROUND AT |
| | | |

| FASTEST TIME | FASTEST LAP |
|---|---|
| | |
| | |
| | |

## FURTHER OBSERVATIONS

| | | |
|---|---|---|
| RACE EVENT | | DATE |
| LOCATION | | |
| ORGANISER | | / / |

| | | | |
|---|---|---|---|
| RACE NUMBER IN ROW | | CURRENT STANDING | |
| WEATHER CONDITION | | TIME OF YEAR | |
| TIRES USED | | RACE START | |
| | | RACE FINISH | |
| CIRCUIT NAME | | RACING CLASS | |

| TYRE PRESSURE | | START POSITION |
|---|---|---|
| FRONT | REAR | |
| | | LAPS COMPLETED |
| | | |
| | | 1ST ROUND AT |
| | | |

| FASTEST TIME | FASTEST LAP |
|---|---|
| | |
| | |
| | |

FURTHER OBSERVATIONS

| | | |
|---|---|---|
| RACE EVENT | | DATE |
| LOCATION | | |
| ORGANISER | | / / |

| | | | |
|---|---|---|---|
| RACE NUMBER IN ROW | | CURRENT STANDING | |
| WEATHER CONDITION | | TIME OF YEAR | |
| TIRES USED | | RACE START | |
| | | RACE FINISH | |
| CIRCUIT NAME | | RACING CLASS | |

| TYRE PRESSURE | |
|---|---|
| FRONT | REAR |
| | |
| | |

**START POSITION**

**LAPS COMPLETED**

**1ST ROUND AT**

| FASTEST TIME | FASTEST LAP |
|---|---|
| | |
| | |
| | |

**FURTHER OBSERVATIONS**

| | | |
|---|---|---|
| RACE EVENT | | DATE |
| LOCATION | | |
| ORGANISER | | / / |

| | | | |
|---|---|---|---|
| RACE NUMBER IN ROW | | CURRENT STANDING | |
| WEATHER CONDITION | ☼ ☁ ❄ ⛈ ≋ | TIME OF YEAR | 🌷 ☼ 🍂 ❄ |
| TIRES USED | | RACE START / RACE FINISH | 🕒 |
| CIRCUIT NAME | 🏁 | RACING CLASS | |

| TYRE PRESSURE | | START POSITION |
|---|---|---|
| FRONT | REAR | |
| | | LAPS COMPLETED |
| | | |
| | | 1ST ROUND AT |
| | | |

| FASTEST TIME | FASTEST LAP |
|---|---|
| | |
| | |
| | |

## FURTHER OBSERVATIONS

| RACE EVENT | | DATE |
|---|---|---|
| LOCATION | | / / |
| ORGANISER | | |

| | | | |
|---|---|---|---|
| RACE NUMBER IN ROW | | CURRENT STANDING | |
| WEATHER CONDITION | | TIME OF YEAR | |
| TIRES USED | | RACE START / RACE FINISH | |
| CIRCUIT NAME | | RACING CLASS | |

## TYRE PRESSURE

| FRONT | REAR |
|---|---|
| | |
| | |

## START POSITION

## LAPS COMPLETED

## 1ST ROUND AT

## FASTEST TIME

## FASTEST LAP

## FURTHER OBSERVATIONS

| | | |
|---|---|---|
| RACE EVENT | | DATE |
| LOCATION | | |
| ORGANISER | | / / |

| | | | |
|---|---|---|---|
| RACE NUMBER IN ROW | | CURRENT STANDING | |
| WEATHER CONDITION | ☼ ☁ ❄ ⚡ ≋ | TIME OF YEAR | ❀ ☼ ❦ ❄ |
| TIRES USED | | RACE START<br>RACE FINISH | 🕒 |
| CIRCUIT NAME | 🏁 | RACING CLASS | |

## TYRE PRESSURE

| FRONT | REAR |
|---|---|
| | |
| | |

## START POSITION

## LAPS COMPLETED

## 1ST ROUND AT

## FASTEST TIME

## FASTEST LAP

## FURTHER OBSERVATIONS

| | | |
|---|---|---|
| RACE EVENT | | DATE |
| LOCATION | | |
| ORGANISER | | / / |

| | | | |
|---|---|---|---|
| RACE NUMBER IN ROW | | CURRENT STANDING | |
| WEATHER CONDITION | ☼ ☁ ❄ ⛈ ≋ | TIME OF YEAR | ✿ ☼ ❦ ❄ |
| TIRES USED | | RACE START<br>RACE FINISH | ◷ |
| CIRCUIT NAME | ⚑ | RACING CLASS | |

| TYRE PRESSURE | | START POSITION |
|---|---|---|
| FRONT | REAR | |
| | | LAPS COMPLETED |
| | | |
| | | 1ST ROUND AT |
| | | |

| FASTEST TIME | FASTEST LAP |
|---|---|
| | |
| | |
| | |

FURTHER OBSERVATIONS

# *Thanks For Reading!*

Just a quick message to thank you so much for picking up one of our books! Our sincere hope is that this book has given you the value we always look to provide, and hope we can continue to produce quality books that will in anyway contribute to a better quality of life for our readers.

We are a small independent publisher based in London, UK and we work with talented authors from around the world, who dedicate every ounce of their effort to craft these memorable books for your reading pleasure.

The author of this title would love to hear about your experience with the book, and your review will go a long way to provide them with the insight and encouragement they need to keep creating the kind of books you want to read.

**Your Opinion Makes a Real Difference.**

If you want to let us know what you thought about the book, please visit the Amazon website and give us your review. We read every single review, no matter how long or short!

HAPPY READING!

Made in the USA
Monee, IL
04 November 2026

67379706R00069